CANCER AND CBD OIL

Understanding The Benefits Of Cannabis And
Medical Marijuana

Table of Contents

other than to increase dietary intake levels of specific nutrients is neither implied nor advocated by the author.

The results reported may not occur in all individuals. This is a comprehensive limitation of liability for damages arising out of or in connection with the use of this book of any kind, including (without limitation) compensatory, direct, indirect or consequential damages, loss of income, profit, loss or damage to personas for property claims by third parties, The below is not intended to diagnose, treat, cure or prevent any disease.

The below information is provided for educational and information purpose only and is not intended to be a substitute for a health care provider's consultation. The information in this book should not be relied upon to suggest a course of treatment for any particular individual. It should not be used in place of a consultation with your physician or other qualified healthcare professional.

INTRODUCTION

The plant Cannabis sativa has been used in medicinal practice for thousands of years. Cannabis is a complex plant with over 400 chemical entities of which more than 60 of them are cannabinoid compounds, some of them with opposing effects. The pharmacologically active constituents of the plant are termed cannabinoids, phytocannabinoids (cannabinoids derived from the plant), synthetic cannabinoids (artificial compounds with cannabinomimetic effects), and endocannabinoids (endogenous compounds with cannabinomimetic effects) all of which act together on the human endocannabinoid system (ECS), which regulates various functions in the body.

The roots of Cannabidiol (CBD), specifically the seeds and cannabis oil were used for food in China as early at 6,000 BC. Today, individuals with simple to chronic disease, organic and health conscientious supplement users and alternative medicine seekers are more interested in the health-related properties of the Cannabis compounds in the fight against chronic conditions and debilitating disorders including Acne, Arthritis, Cancer, Crohn's Disease, Epilepsy, Fibromyalgia, Glaucoma, Hepatitis C, HIV/AIDS, Multiple Sclerosis, Nausea, Parkinson's Disease, PTSD - Post Traumatic Stress Disorder, Schizophrenia and everyday Stress are turn to a new found (but ancient) natural medicine in their fight to cure themselves. Those with life-threatening diseases, to those in chronic pain and those with common aliments are seeking the benefits of CBD based Oils and products derived from Cannabis to help alleviate conditions, aid in healing and increase general well being.

In the United States, Cannabis Sativa is a Schedule I substance and its use for recreational or medicinal means is illegal according to Federal law. However, contrary to Federal policy, individual state laws have allowed for medical use of marijuana in 29 states[1] and recreational use in 9 states; including Washington D.C.

Given the evolving policies regarding the medical use of cannabis, physicians are increasingly prompted with questions about its therapeutic role for a variety of disorders. Research, studies and science are now able to prove the medicinal benefits of CBD and THC oil, and finally offer natural alternatives to pharmaceuticals and traditional prescriptions. Consumers and patients both can now fight conditions and aliments by turning to CBD Oil which is natural, plant based, opioid free, healthier and better aligned to an organic chemical free lifestyle.

[1]http://medicalmarijuana.procon.org/view.resource.php?resourceID=000881

CHAPTER ONE: WHAT IS CBD OIL?

Cannabis oil, known as CBD oil, is derived from the seed, stalk and flowers of the cannabis sativa plant and has a significant amount of the compound Cannabidiol. Cannabidiol is one of the more than 111 *active* cannabinoids identified in the cannabis sativa plant. CBD oil is the result of utilizing one of numerous extraction methods to isolate, preserve and maintain the purity of the medicinal resin (oil or residue) that is found on the flowering leaves of the plant.

CBD and CBD oil independently is not psychoactive. This means that it does not change the state of mind of the person who uses it. However, it does appear to produce significant changes in the endocannabinoid system (ECS) system of the human body, and it is proving to have a long list of medical benefits. Most of the CBD used in today's consumer products is found in the least processed form of the cannabis plant, known as hemp. Hemp is different than CBD although derived from the same plant.

For over 40 years researchers have been looking at the potential therapeutic uses of cannabinoids and the different concentration levels along with the interactive properties of CBD and THC, independent and collectively. Until recently, the most well-known compound in the cannabis plant was THC, or Delta-9 Tetrahydrocannabinol. THC is the most active ingredient in Marijuana, which is also found in the Cannabis Sativa plant. Marijuana contains both THC and CBD, but the compounds have different effects. THC is well-known for the mind and body "high" it produces when ingested, such as when smoking the plant or when using oil in edible form, such as for cooking into foods.

CHAPTER TWO: WHAT ARE CANNABINOIDS?

Cannabinoids are naturally occurring compounds found in the Cannabis sativa plant. Over 480 different compounds are present in the plant, but only around 60 are termed *active* cannabinoids. The most well-known among these compounds is Delta-9-Tetrahydrocannabinol (9-THC), which is the main psychoactive ingredient in cannabis. Cannabidiol (CBD) is another predominant and important compound present, which makes up about 40% of the plant resin extract.

For the purpose of this book, we will focus on specific cannabinoids, which are separated into the following subclasses:

Cannabidiol (CBD)

Cannabichromene (CBC)

Cannabigerol (CBG)

Cannabinodiol (CBDL)

Cannabinol (CBN)

Tetrahydrocannabinol (THC)

Differences between Cannabinoids

The main way in which the cannabinoids are differentiated is based on their degree of psycho-activity.
For example, CBG, CBC and CBD are not known to be psychologically active agents whereas THC, CBN and CBDL along with some other cannabinoids are known to have varying degrees of psycho-activity. The most abundant of the cannabinoids is CBD, which is thought to have a wide range of benefits, one being anti-anxiety effects, which are possibly counteracting the psychoactive effects of THC when in the same strain.

When THC is exposed to the air, it becomes oxidized and forms CBN which also interacts with THC to lessen its impact. This is why cannabis that has been left outside of an air-tight container, will has less potency and residual effects when smoked due to the increased CBN to THC ratio.

Effects of Cannabinoids
Cannabinoids exert their effects by interacting with specific cannabinoid receptors present on the surface of cells throughout our body. These receptors are found in different parts of the human central nervous system and the two main types of cannabinoid receptors are referenced as CB1 and CB2.

Laboratory research[2] in 1992, yielded findings of a naturally occurring substance in the brain that binds to CB1, Anandamide. Although discovered decades previous the relation to the CB1 receptors was not understood. This cannabinoid-like chemical and others that were later discovered are referred to as endocannabinoids.

The effects of cannabinoids depends on the area of the brain which are involved. Effects on the limbic system may alter the memory, cognition and psychomotor performance; effects on the mesolimbic pathway may affect the reward and pleasure responses and pain perception may also be altered.

Δ-9-tetrahydrocannabinol (THC)

[2]https://www.researchgate.net/publication/6126760_Discovery _and_Isolation_of_Anandamide_and_Other_Endocannabinoid s

Primary Cannabinoids in Cannabis

THC

THC is the abbreviation for both Delta(9)-Tetrahydrocannabinol and Delta(8)-Tetrahydrocannabinol, which is the lesser psychoactive THC cannabinoid. As most cannabis lovers probably know, THC is the primary psychoactive compound in marijuana. The list of medical conditions of which THC is found to have benefit is rapidly growing, so for purpose of this book we will focus on a few of the notable conditions with published studies and research:

- Alzheimer's Disease
- Neuropathic pain
- Multiple Sclerosis
- Parkinson's Disease
- PTSD
- Cancer
- Crohn's Disease
- Chronic (pain) relief

Some resources claim that Delta(8) may have neuroprotective and anti-anxiety properties, making it an interesting companion to the more notorious psychoactive. However, more research is needed to confirm just how this particular compound acts inside the body. Here are a couple of additional benefits you can expect from Delta(8)-THC:

- Anti-Anxiety
- Appetite Stimulator
- Pain Relief

- Neuroprotection
- Anti-Nausea

CBD

CBD, which is short for Cannabidiol, is the second most famous cannabinoid. Like THC, the list of medical benefits of this cannabinoid just keeps getting longer. Unlike THC, CBD is non-psychoactive. It's also now legal in more states than its more controversial counterpart and available in a large variety of products and forms. Once mainstream media learned that CBD had medical value, a whirlwind of cannabis research ensued. Here are a just a handful of conditions CBD can treat:

- Acne
- Anxiety
- Arthritis
- Cancer
- Depression
- Diabetes
- Epilepsy
- Fibromyalgia
- Glaucoma
- Hepatitis C
- Psychotic Disorders
- Chronic Pain
- Multiple Sclerosis
- Nausea
- Parkinson's Disease
- PTSD Post Traumatic Stress Disorder
- Schizophrenia
- Stress

CBC

Cannabichromene (CBC) is an abundant naturally-occurring phytocannabinoid, and is thought to be the second most abundant cannabinoid in cannabis. CBC has been shown to produce anti-nociceptive (painkilling) and anti-inflammatory effects in studies. CBC shares the same molecular formula as THC and CBD: $C_{21}H_{30}O_2$. Although many cannabinoids share the same formula, the atoms within the molecule are arranged in slightly different ways. In some strains, CBC may even take dominance over CBD. Similar to CBD, Cannabichromene is non-psychoactive.

Here are benefits of CBC rich strains:

- Encourages Brain Growth
- Anti-Inflammatory
- Anti-Depressant
- Anti-Bacterial and Anti-Fungual
- Pain Relief

CBN

Cannabinol, emerges when the dried flower has gone a bit stale. THCa breaks down into this compound over time and results in a mildly different compound. CBN's most pronounced, characterizing attribute is its sedative effect, and studies are showing, 5mg of CBN is as effective as 10mg dose of diazepam, a mild pharmaceutical sedative. If you leave some bud out sitting out in the open air for too long, you'll eventually have a product with larger amounts of CBN. CBN's studied benefits include:

- Appetite Stimulant

- Anti-Inflammatory
- Anti-Bacterial
- Pain reliever
- Anti-Asthmatic
- Anti-Insomnia
- Potential Medication for Glaucoma

CBG

Cannabigerol (CBG) is the building block for many compounds in marijuana, including THC and CBD. This cannabinoid is found early on in the growth cycle, which makes it somewhat difficult to find in large quantities. In fact, THC, CBD and many other cannabinoids all begin as CBG. CBG is non-psychoactive and can be thought of as the "stem cell" or "parent" of other cannabinoids. After being synthesized, CBG is quickly converted to other cannabinoids through natural processes that occur within the cannabis plant. This explains the low CBG content of most cannabis strains. Research and studies are finding CBG benefit in supporting:

- Anxiety and Pain
- Glaucoma
- Anti-Inflammation
- Skin Conditions
- Digestive Conditions
- Anti-Septic
- Neuroprotection
- Cancer
- Huntington's Disease

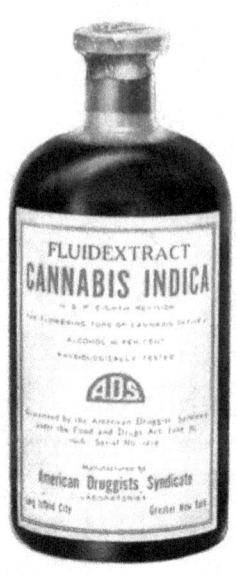

CHAPTER THREE: HISTORY AND MISCONCEPTIONS OF CBD OIL

Cannabis (Medical marijuana) has long been used to treat a variety of ailments and conditions since ancient times. However, CBD oils and related products are now a prevalent topic in today's medical field with continued discoveries of the profound effect on children and adults with debilitating and chronic disease.

Using CBD dates back to the 19th century and is noted Queen Victoria would use cannabis to alleviate menstrual cramp pain. CBD oils and related products until recently have always played second to its fellow cannabinoid THC. In the 1980s, studies hinted that CBD could alleviate certain types of pain, anxiety and nausea, but still did not get much attention until the 90s.

In 1998, a medical company called GW Pharmaceuticals based in England began to cultivate cannabis specifically for medical trials. Their aim was to develop a concise and consistent plan for extracting CBD. Geoffery Guy[3], one of the company's founders, staunchly believed that cannabis plants rich in CBD would be used as medicine. This research would lead to scientific studies conducted by the International Cannabinoid Research Society[4], Society of Cannabis Clinicians[5], and International Association for Cannabinoid Medicine[6]. Early studies showed that CBD lessened anxiety and reduced the frequency and severity of seizures. This turned heads in the medical community, and cannabis strains began to be cultivated with extremely low levels of THC and high levels of CBD.

In the 2010s, the public began to see what a profound effect CBD oil could have treating a variety of life-threatening ailments, especially in children. One of the most encouraging and perception changing stories is that of Charlotte's Web[7], which is named after Charlotte Figi[8], born October 18, 2006. Her story has led to her being described as "the girl who is changing medical marijuana laws across America." Her parents and physicians say she experienced a reduction of her

---

[3] https://www.gwpharm.com/about-us/board-directors
[4] http://icrs.co/#About
[5] http://cannabisclinicians.org/
[6]https://www.cannabis-med.org/index.php?tpl=page&id=73&lng=en

[7] https://en.wikipedia.org/wiki/Charlotte%27s_web_(cannabis)
[8] https://en.wikipedia.org/wiki/Charlotte%27s_web_(cannabis)

epileptic seizures brought on by Dravet syndrome after her first dose of medical marijuana at five years of age. Her usage of Charlotte's Web CBD Oil was first featured in the 2013 CNN documentary "Weed". Media coverage increased demand for Charlotte's Web and similar products high in CBD, which has been used to treat epilepsy in toddlers and children.

In the modern era, advances in chemistry have allowed people to consider both raw CBD oil, closer to its minimally filtered historical form, and more refined extractions that increase the levels of CBD. Moreover, developments in technology and extraction have created a variety of formulas and products like dermal skin patches, lotions, balms, and a wide variety of edibles. Today, the options in CBD oil supplements are so vast the first farmers could have never dreamed of, but of course, this also requires that each person carefully investigate the supplements that are right for them condition and desired benefit.

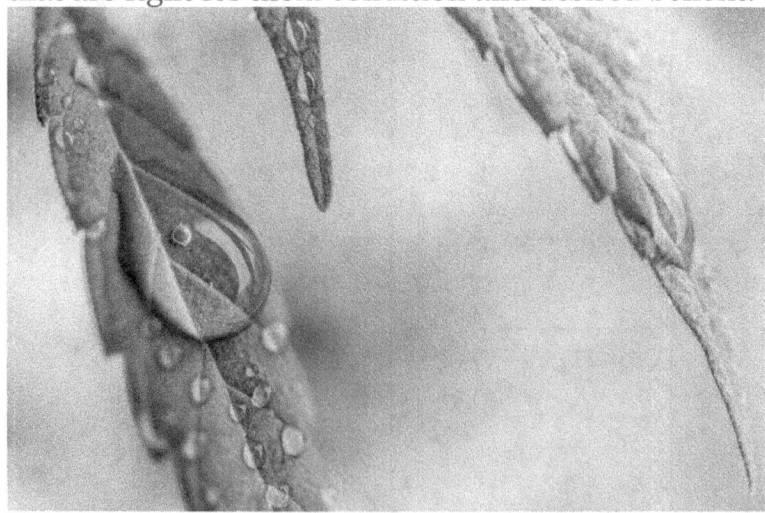

Misconceptions

Cannabis has gotten a wave of new momentum following Dr. Sanjay Gupta's documentary regarding the medicinal properties and benefits of cannabis. Cannabidiol - or CBD - has been given the title of being medicinal because it treats without giving a person the signature high cannabis has been known for. THC or Tetrahydrocannabinol is that cannabinoid or the non-medical type which has gotten much more attention in the past. Both of these play important roles in medicinal marijuana treatments, but there are several misconceptions ,which need to be cleared up.

From medical marijuana centers, to cutting edge medical teams, and even across major news outlets, it's a buzzword that is only growing in popularity and interest. While CBD does not get you "high," like its cousin THC, it has certainly piqued the interest of everyone from seasoned marijuana growers to industrial hemp farmers to esteemed medical teams. With great buzz, comes information and a myriad of "facts" claiming that it is a cancer-curing, seizure-stopping miracle oil. While there are certainly a large number of cases cited with patients who CBD and CBD Oils have helped change or improve their conditions significantly, it is imperative to separate what we know as scientific evidence from fallacies about the cannabinoid. Plenty of start-ups and e-commerce retailers have jumped on the CBD bandwagon, touting CBD derived from industrial hemp as the next big thing, a miracle oil that can shrink tumors, reduce or stop seizures, and ease chronic pain without making people feel "stoned." But along with a growing awareness of Cannabidiol as a potential health aid there has been a proliferation of misconceptions about CBD. Here are six of them.

20

"CBD is medical. THC is recreational."

Often people say they are seeking "CBD, the medical part" of the plant, "not THC, the recreational part" that gets you high. Actually, THC, "The High Causer" has a growing list of therapeutic benefits and properties. Scientists at the Scripps Research Center in San Diego reported that THC inhibits an enzyme implicated in the formation of beta-alkaloid plague, the hallmark of Alzheimer's-related dementia. The federal government recognizes single-molecule THC (Marinol) a synthetic man-made THC structurally similar to its natural counterpart but dissolved in sesame oil, as an anti-nausea compound and appetite booster, deeming it a Schedule III drug, a category reserved for medicinal substances with little abuse potential. Ironically, the whole marijuana plant, the only natural source of THC, continues to be classified as a dangerous Schedule I drug with no medical value.

"THC is the bad cannabinoid. CBD is the good cannabinoid."

The Cannabis Hater strategic retreat: Give ground on CBD while demonizing THC. Diehard marijuana haters are exploiting the good news about CBD to further stigmatize THC in cannabis, casting Tetrahydrocannabinol as EVIL, whereas CBD is framed as the GOOD. Why? Because CBD doesn't make you high like THC does.

"CBD is most effective without THC."

THC and CBD are the power couple of the cannabis compound as they work well together. Scientific studies have established that CBD and THC interact in synergy to enhance each others therapeutic effects. British researchers have shown that CBD potentiates THC's anti-inflammatory properties in an animal model of colitis. We encourage each person to discover what works for them and of course if you live in a state where Marijuana is illegal for recreational use then you might want to consider moving to a state like Colorado who has legalized adult use.

"Single-molecule pharmaceuticals are superior to 'crude' whole plant medicinal."

According to the US Government, specific components of THC and CBD have medical value, but the plant itself does not have medical value. Uncle Sam's single-molecule blinders reflect a clear bias that privileges large pharmaceutical products. Single-molecule medicine is the predominant corporate way, the FDA-approved way, but it's not the only way, and it's not necessarily the optimal way to benefit from cannabis therapeutics.

"It matters where CBD comes from."

People think it often takes large amounts of industrial hemp to extract a small amount of CBD but that's simply not the case anymore as many of these industrial hemp plants have been bred specifically to have a high CBD content. The extraction method can also affect how much CBD is extracted per plant as well as its potency. Almost all hemp derived CBD oil is made using CO_2 extraction as opposed to most marijuana extractions that used a harsh chemical solvent. While medical marijuana legislation is opening up and more and more states are planting crops for medical and commercial purposes, it is by no means the only means to extract CBD. In fact, these hemp crops tend to produce almost identical THC:CBD ratios as medical marijuana.

"CBD is legal throughout the United States"

CBD products are illegal in the United States based on Federal Law, if the product contains any amount of THC. A product that contains THC, even in small amounts, is considered Marijuana under federal law and is illegal. When it comes to CBD products such as tinctures, lotions, balms, and beyond, the cannabinoid can be extracted from leaves, flowers, and the stalk, similarly to industrial hemp. While legislation is in motion to ease up on industrial hemp cultivation, there are only a few states that allow the cultivation for commercial, research or pilot programs.

CHAPTER FOUR: HOW CBD OIL IS MADE

Cannabidiol (CBD) oil is derived from the Cannabis sativa plant and extracted specifically from the high-resin glands or *trichomes*[9] found mainly on the plant's odiferous female flowers (the buds) and to a lesser extent on the leaves. There are also the smaller trichomes, which dot the stalk of the hemp plant, but these contain hardly any resin. Non-glandular hairs shaped like tiny inverted commas also cover the plant's surface.

One of the original creation methods and the man who re-invented CBD oil is Canadian, Rick Simpson. Rick Simpson oil (RSO) is highly-concentrated cannabis oil extracted from female cannabis plants that contain at least 20 percent THC or more. Simpson favors high-THC and low-CBD content when treating maladies. RSO was created when he recalled a study published in the Journal of the National Cancer Institute[10] stating that THC was found to kill cancer in mice. Simpson proceeded to apply the oil topically to recently identified cancerous spots on his neck and face and covered them with bandages. In just four days, the skin underneath was healthy and pink. He was healing. Excited that he had discovered a cure to his cancer, Simpson began helping others heal, spreading the good word ever since.

Despite encountering opposition from his own doctor, local authorities, pharmaceutical companies, government health agencies, and the United Nations, Simpson has not only healed his own ailments, he's

---

[9] https://en.wikipedia.org/wiki/Trichome
[10] https://www.cancer.gov/about-cancer/treatment/cam/patient/cannabis-pdq

also successfully treated over 5,000 patients free of charge for all kinds of conditions, including cancer, HIV/AIDS, insomnia, diabetes, depression, osteoporosis, asthma, and more. He even cured his mother's psoriasis.

The active CBD compound in the flower can be extracted, purified and concentrated using a limited number of precise extraction and production techniques. When oil of a higher potency and concentration is derived from the glands it can be used in lower dosage. The higher purity and lower dosage can save money as CBD oil is an expensive product to produce at the highest quality and desired compound blend.

Industrial production of CBD oil is done by combining the cannabinoid rich plant with other compounds like CO_2, butane, ethanol or olive oil, of which can leave their residues in the final product. The most important part of CBD oil extraction is selection of right plant for the type of oil extraction. Typically cost of the final product depends on its potency and purity, origin, heritage, along with the grow process that was used. Oil is in liquid form it can be consumed orally without smoking and can be easily administered to children and adults for whom it is prescribed as medication.

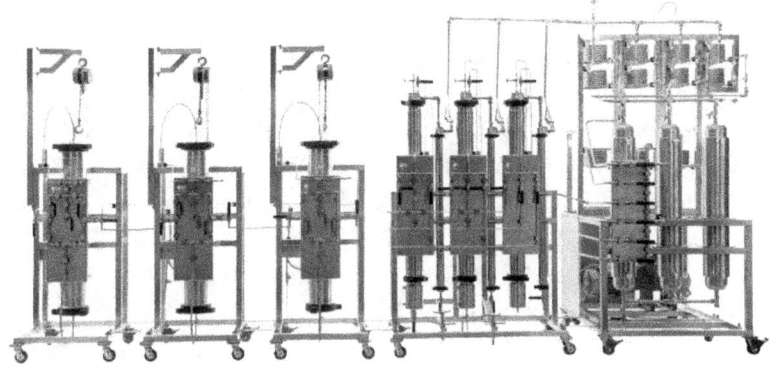

Why is Cannabis extraction performed?

Cannabis extraction can be performed for a variety of reasons, as well as in variety of ways. The most widespread reason, of course, is to obtain potent extracts high potency resin that can be used in the medical marijuana industry to treat patients dealing with a variety of illnesses and symptoms. On a more academic level, cannabis extractions are also performed purely for research purposes in an effort to understand the chemistry of the plant and make scientific inquiries into certain claims.

Just as other compounds might be available in various solid or liquid forms, the effects of cannabis can be achieved in a variety of extract forms. The following is a list of just a few of the most common cannabis extraction methods:

SuperCritical Co2

CO_2 extraction machines essentially freeze and compress CO_2 gas into a "supercritical" cold liquid state. Carbon dioxide is considered a "supercritical fluid" that becomes a liquid under pressure. In extractions, CO2 leaves no residue, which makes it popular for use in extraction processes across a variety of industries. In this kind of cannabis extraction, supercritical CO2 is put into a pressurized container with the cannabis material. The mixture is then put through a filter to separate the CO2 from the cannabis and, when the pressure in the system is released, the CO2 evaporates back into a gas.

Referred to as "subcritical" or "supercritical" CO2 method, it uses carbon dioxide under different pressures to extract the medicinal oil. By pushing Co2 through the plant at high pressures and low temperatures, CBD can be extracted in its purest form. The CO2 method is considered the most scientific method and this is also among the cleanest techniques of extraction.

This process is often thought of as the best and safest as it cleanly extracts CBD, removing substance like chlorophyll and leaving no residue. CBD oil extracted in this way has a cleaner taste and this is among the most expensive techniques of extraction due to use of hi-tech equipment that must be operated by trained professionals. The advantage of this technique is that the end product is the purest form of CBD oil, which is highly potent and free of chlorophyll and other potential harmful toxins. However, if the heat used during extraction is too high it can damage Terpenes in the oil that have therapeutic benefits and provide flavor and essence to the oil strain.

Distillation

Distillation is probably the most common technique; short path distillation exposes the cannabis oil to heat and vacuum in order to separate a variety of compounds from the extract (cannabinoids and Terpenes, in particular). Using an appropriate laboratory setup, heat is first introduced to the extract to evaporate the cannabinoids and Terpenes. Then, the vapor enters a condenser tube, through which the cannabinoids travel and condense into the recipient flask. This final product does not need to be winterized, as the waxes and fats cannot vaporize and, therefore, remain in the first flask of unprocessed concentrate. This method generally produces the highest concentrations of cannabinoid molecules.

Carrier Oil

The carrier oil extraction method is regarded as most inexpensive method of extracting CBD oil and is recommended by Dr. Arno Hazekamp[11], director of phytochemical research at Bedrocan BV that supplies it to Dutch Health Ministry. CBD oil extracted by this method will contain a healthy dose of Omega rich acids and minimalistic chemical residues in the pure oil. Usually hemp seed oil or olive oil is used as a carrier in these methods as this is the most effective way to extract resin from the plants and flowers. The only drawback is the short shelf life of oil extracted in this manner though it is highly effective when taken orally or applied topically on the skin.

Live Resin Extraction

Most commonly, cannabis is dead and dried before it is processed. In Live Resin Extraction, the live plant is

---

[11] https://www.researchgate.net/profile/Arno_Hazekamp

frozen (such as with liquid nitrogen) right after it is harvested in order to perform extraction. Using a closed-loop hydrocarbon extraction machine this process creates a resin product. The live resin extraction method generally gives a flavor that is more true to the original strain and uses the living plant. This is also one of the most expensive methods to perform because most of the plant's terpenes (essential oils) in the product resulting in a high potency oil.

Terpene Isolation

Terpenes are hydrocarbon oils that are responsible for fragrance in a variety of plants. A wide variety of plant essential oils including those with pine or citrus scents are comprised primarily of terpenes. In the case of cannabis, some experts believe that the terpenes of the plant are equally as responsible as THC for the high experienced during use.

Solvents

Extraction with solvents is a relatively inexpensive method and typically preferred by small-scale producers of CBD and THC oils. During extraction this method uses solvents like butane, ethanol and alcohol derived from grains. This method has several disadvantages the worst of which is potential of explosion while the second is leftover residue of these solvents. Scientists and doctors advice against the use of this method as it can make the end product unsafe for medical use and also can make existing medical condition much worse. When there are unsafe residues in CBD oil it reduces healing powers and can even compromise health of patients.

CHAPTER FIVE: HOW CBD WORKS

All cannabinoids, including CBD, attach themselves to certain receptors in the body to produce their effects. The human body produces certain cannabinoids on its own. It has two receptors for cannabinoids, called CB1 receptors and CB2 receptors.

The CB1 receptor was discovered in 1990, while CB2 was uncovered shortly thereafter in 1993 by a research group at Cambridge University. One source claims that these two receptor types employ significantly different signaling mechanisms. It is known that they are expressed in vastly different ways, including their appearance in various parts of the body (different regions of the endocannabinoid system).

CB1 receptors are present in very high levels in several brain regions and in lower amounts throughout the body and nervous system. The CB1 receptors, found predominantly in the *cerebellum* and *neocortex* regions of the brain deal with motor coordination and initiation of movement, pain, emotions and mood, thinking, appetite, and memories, among others. THC attaches to these receptors and help mediate many of the psychoactive effects of cannabinoids.

THC has been shown to possess a high binding affinity with CB1 receptors in the brain, central nervous system, connective tissues, gonads, glands, and related organs. This is one reason that consumption of cannabis oils from strains containing a high amount of THC result in a relatively potent effect, giving patients significant relief from pain, nausea, or depression while delivering a strong euphoria to lifestyle users. Those undergoing chemotherapy and patients suffering conditions involving inflammation, like arthritis and lupus, gain significant efficacy.

Think of it like an electrical plug connecting to a wall

socket. A THC molecule is perfectly shaped to connect with CB1 receptors. When that connection happens, THC activates, or stimulates, those CB1 receptors. Researchers call THC a CB1 receptor agonist, which means THC works to activate those CB1 receptors. THC partially mimics a naturally produced neurotransmitter known as anandamide, aka "the bliss molecule." Anandamide is an endocannabinoid, which activates CB1 receptors.

When we're talking about cannabis and psycho-activity, we're dealing exclusively with CB1 receptors, which are concentrated in the brain and the central nervous system. The difference between CBD vs THC comes down to a basic difference in how each one interacts with the cannabinoid (CB1) receptor. THC binds well with CB1 cannabinoid receptors. That's where the two diverge.

CB2 receptors, on the contrary, are located throughout the immune system and related organs, like the tissues of the spleen, tonsils, and thymus gland. They are also common in the brain, although they do not appear as densely as CB1 sites and are found on different types of cells.

CB2 sites are also found in greater concentrations (density) throughout the gastrointestinal system, where they modulate intestinal inflammatory response. This is why sufferers of Crohn's disease and IBS gain such great relief from cannabis medicine. It is also a powerful example of how the endocannabinoid system, when supplemented by external cannabinoids (such as from cannabis), can provide such powerful and long-lasting relief for patients of diseases like Crohn's. Cannabis and CBD oil has been shown to have such great efficacy for this condition that, in nearly half of cases, the medicine puts the disease into full remission. The author can personally affirm to the medicinal benefits of CBD and THC oil used in the treatment, remission and chronic pain relief of Crohn's disease as diagnosed in 1983.

Serotonin Receptor

At high concentrations, CBD directly activates the 5-HT1A (hydroxytryptamine) serotonin receptor, thereby conferring an anti-depressant effect. This G-coupled protein receptor is implicated in a range of biological and neurological processes, including (but not limited to) anxiety, addiction, appetite, sleep, pain perception, nausea and vomiting. 5-HT1A is a member of the family of 5-HT receptors, which are activated by the neurotransmitter serotonin. Found in both the central and peripheral nervous systems, 5-HT receptors trigger various intracellular cascades of chemical messages to produce either an excitatory or inhibitory response, depending on the chemical context of the message. CBDA, Cannabdiolic acid, the raw, unheated version of CBD that is present in the cannabis plant, also has a strong affinity for the 5-HT1A receptor (even more so than CBD).

Both CBD and CBDA trigger an inhibitory response that slows down 5-HT1A signaling. In comparison, LSD, mescaline, magic mushrooms, and several other hallucinogenic drugs activate the 5-HT2A receptor, which produces an excitatory response.

Vanilloid Receptors

TRPV is the technical abbreviation for "Transient Receptor Potential Cation" Channel Subfamily V." TRPV1 is one of several dozen TRP (pronounced "trip") receptor variants or subfamilies that mediate the effects of a wide range of medicinal herbs.

CBD directly interacts with various ion channels to confer a therapeutic effect. CBD, for example, binds to TRPV1 receptors, which also function as ion channels.

TRPV1 is known to mediate pain perception, inflammation and body temperature. Scientists also refer to TRPV1 as a "Vanilloid receptor," named after the flavorful vanilla bean. Vanilla contains Eugenol, an essential oil that has antiseptic and analgesic properties; it also helps to unclog blood vessels.

CBD is a TRPV1 "agonist" or stimulant. This is likely one of the reasons why CBD-rich cannabis is an effective treatment for neuropathic pain.

GPR55 Orphan Receptor

Whereas Cannabidiol directly activates the 5-HT1A serotonin receptor and several TRPV ion channels, some studies indicate that CBD functions as an antagonist that blocks, or deactivates, another G protein-coupled receptor known as GPR55.

GPR55 has been dubbed an "orphan receptor" because scientists are still not sure if it belongs to a larger family of receptors. GPR55 is widely expressed in the brain, especially in the cerebellum. It is involved in modulating blood pressure and bone density, among other physiological processes. GPR55 promotes osteoclast cell function, which facilitates bone reabsorption. Overactive GPR55 receptor signaling is associated with osteoporosis. GPR55, when activated, also promotes cancer cell proliferation, according to a 2010 study by researchers at the Chinese Academy of Sciences in Shanghai. This receptor is expressed in various types of cancer.

CBD is a GPR55 antagonist, as University of Aberdeen scientist Ruth Ross disclosed at the 2010 conference of the International Cannabinoid Research Society in Lund, Sweden. By blocking GPR55 signaling, CBD may act to decrease both bone reabsorption and cancer cell proliferation.

PPARs - nuclear receptor

CBD also exerts an anti-cancer effect by activating PPARs [peroxisome proliferator activated receptors] that are situated on the surface of the cell's nucleus. Activation of the receptor known as PPAR-gamma has an anti-proliferative effect as well as an ability to induce tumor regression in human lung cancer cell lines. PPAR-gamma activation degrades amyloid-beta plague, a key molecule linked to the development of Alzheimer's disease. This is one of the reasons why Cannabidiol, a PPAR-gamma agonist, may be a useful remedy for Alzheimer's patients.

PPAR receptors also regulate genes that are involved in energy homeostasis, lipid uptake, insulin sensitivity, and other metabolic functions. Diabetics, accordingly, may benefit from a CBD-rich treatment regimen. CBD also exerts an anti-cancer effect by activating PPARs on the surface of the cell's nucleus.

CHAPTER SIX: CHOOSING THE BEST CDB
OIL FOR YOUR NEEDS

Trying to find the best CBD oil can be very tough especially when you are just starting your research. Honestly, if you want to choose the best Cannabidiol or hemp oil and its products; you'll have to go through several channels and sometimes a variety of options in order to get correct information, strain and balance to help with the condition you are trying to alleviate.

Many people still confuse Cannabidiol (CBD) with Tetrahydrocannabinol (THC), the major psychoactive ingredient found in cannabis. Furthermore, CBD and hemp oil products are often confused to their origin, and medicinal properties. A large variety of CBD specific products have been legalized in a number of states because they do not give the high of THC, which many people run away from due to its psychological effects.

Many cannabis enthusiasts believe that the natural hemp extracts are the next generational as well as the next revolution in nutritional supplements. They believe that there are wider ranges of products that can help to deal with diseases fighting humanity in a natural and healthy way.

Things To Consider Before Choosing Your CBD Oil

CBD oil comes in various forms and products types. Choosing CBD oil can be overwhelming considering the fact that there are significant numbers of CBD products and brands in the market. Remember, CBD and THC work best together, enhancing each other's therapeutic benefits. For maximum therapeutic impact, choose products that include CBD and THC.

So with all these varieties in the market of today, which CBD oil is the best choice for you? The first step lies in knowing the way to compare products that are similar and differentiate those that seem identical. Therefore, to make an informed selection, these areas should be given adequate consideration:

CBD Oil Strength

Concentration of CBD in the product is the determining factor when choosing CBD oil products. Although the concentration can be misleading, it's the primary consideration when looking at edibles and digestible products.

Percent volume of CBD in the product - This property listing should be on the label and expressed as the percentage of the total volume of the product. It usually ranges from 0.1 to 0.26 in percentage. The concentration chosen would be dependent the amount of CBD you want to start with as well as the kind of product you would be sourcing it from.

Purity

It's important to consider what other property or thing your CBD oil has. The possibility of including preservation, additives, and solvents cannot just be ignored. Since CBD oil is derived from hemp - stalk, leaves, and flowers; and because farming methods are diverse, the possibility of CBD products contain pesticides, chemical fertilizers, and herbicides is 100% real. To avoid these unwanted chemicals and materials, seek products that are from natural, organic or well-tested synthetic industrial hemp sources.

Transparency

For you to determine the purity level of the CBD oil, the original producers must be ready to certify lab analyses for every step of the product. This will help to show the concentration of CBD in the product as well as prove that the product is free from pesticides and other harsh chemicals. Look for products that are tested for consistency, and verified as free of mold, bacteria, pesticides, solvent residues, and other contaminants. Avoid products extracted with toxic solvents like BHO, propane, hexane or other hydrocarbons. Solvent residues are especially dangerous for immune-compromised patients. Look for products that entail a safer method of extraction like supercritical CO_2.

Price

The price of any product is influenced by its quality and purity. To produce high-quality CBD oil, a substantial amount of hemp is needed and an absolute refinement process is required as well. Select products with quality ingredients. No corn syrup, GMOs, trans-fats, and artificial additives. So, it's understandable that the purer and more concentrated the oil is; the higher the price will likely be.

This is the core reason why it is very important to look for a reputable company that will satisfy your CBD needs. Don't settle for poor quality oil just to save a few dollars, as most likely; you won't be getting any value.

SECTION II:
CANNABINOIDS AND CANCER

CHAPTER SEVEN: WHAT IS CANCER?

For over 20 years there has been research conducted to see how cannabinoids could affect cancer. The results have been astounding. In order to more profoundly understand the results, though, we need to deepen our conception of what cancer is. Generally, we know that cancer involves malignant tumors or malignant cell proliferation inside our bodies, and we know that cancer can be and often is lethal to the patient who has been diagnosed with it, but beyond the fear we do not often know much else. Let us shift our thinking from anxious dread to a mindset of seeking understanding, because with understanding comes the power to overcome.

Cancer occurs when cells become malignant and begin to rapidly multiple without participating in apoptosis, which is the "programmed cell death" that is part of the life cycle of cells and is useful for cleaning up old, no longer usable cells. This means that they build into a mass of non-functioning cells that take over where the old, functional cells were, causing the organ on which they form to malfunction.

Gene Malfunction: The Molecular Cause Of Cancer

Cancer begins as part of one or more of three processes: the conversion of proto-oncogenes to cancerous genes, the malfunction of tumor suppressor genes, and the malfunction of DNA repair genes. Let us look at each of these in a little more detail.

The *proto-oncogenes* are genes within the cell's DNA that control cell growth and cell division.

Sometimes, through various means, these genes are altered, and depending on how they are altered, they can become oncogenes. Oncogenes, or cancer-causing genes, let cells grow and pass under the radar of the immune system to survive when they should not. That is, they continue to exist, grow, and multiply when normally they would be marked by the immune system for destruction.

The **tumor suppressor genes** are also involved, many times, in causing cancerous tumors to grow. These genes are involved in cell growth and cell division, like the proto-oncogenes. Normally they "suppress" cancerous cells' ability to grow and divide in an uncontrolled manner, but when the tumor suppressor gene is altered, it begins to allow cancerous growth and cell division.

DNA repair genes are essential to the human body's ability to fight cancer. Cancer begins with damaged DNA and mutated cells. DNA repair genes help to repair the damaged DNA so that the cell becomes normal and functional. However, sometimes the DNA repair gene itself mutates, inhibiting its ability to repair another mutated piece of DNA. Mutations within the DNA repair genes therefore have a double effect of causing the cell to grow improperly *and* being unable to restore the proper DNA in other strands.

Between these three mutations and malfunctions of genes, scientists can explain the development of many types of cancer. Research is such that nowadays, scientists have even pinpointed which mutations are involved in various types of cancer.

Processes Of Cancer Development

Cancer grows and spreads in four ways: proliferation, metastasis, angio-genesis, and lack of apoptosis. Each of the processes, except metastasis, is fundamental to the growth, survival, and development of an individual cancer, and metastasis is the way in which the cancer spreads to other parts of the body. Let us investigate how each of these processes work, as interrupting these processes is a primary means by which cannabinoids fight cancer.

First, there is ***proliferation*** of the cancer cells. Most cells in the human body grow and divide through a process of meiosis and mitosis a finite number of times before dying through the process of apoptosis. Unlike most cells in the body, cancer cells never cease to undergo meiosis and mitosis, growing and dividing ad infinitum, such that they become a cancerous tumor or another type of cancer like leukemia. Thus, it is not the proliferation itself that is necessarily the issue as much as the endless nature of the proliferation and the lack of an end point that causes cancer to become dangerous. By interrupting this process, the tumor or cancer will be halted from growing.

There is another process in cancer called ***metastasis***, which is the cancer uses to spread to other parts of the body, such as from the breast to the brain or bone or blood. In metastasis, a piece of the tumor, that is, a clump of cells, breaks off from the tumor and travels through the blood or through the lymphatic system to another part of the body, where it attaches and begins to take over that part of the body as well as the original location. If we could interrupt this process, we could stop malignant tumors from spreading and confine them to one area to better be able to treat them and cure them.

Angio-genesis is the means by which cancer cells receive the blood flow they need to survive and grow. "Angio" means blood and "genesis" means creation, and put together they represent the creation of new blood vessels. Tumors need blood flow for the cells to continue to reproduce, as they need the nutrients carried by the blood that are essential to cell growth and they need a flow of blood to take away the toxic cell waste. If we are able to interrupt the process of angio-genesis, then we would be able to not only halt the malignant tumor or cancer's growth but potentially could also begin killing off the cancer cells by cutting them off from the blood supply, which is their life supply.

The final process, an extremely dangerous one, is the continued survival of the cancer cells beyond their normal lifespan. This is called their ***anti-apoptotic*** feature. Apoptosis, as mentioned before, is the "programmed cell death" process that tells a cell when to die. Essentially, it is cell suicide, but it functions to make way for healthy cells when a cell becomes worn out or no longer useful. Apoptosis is the point at which the cell stops proliferating, growing, and dividing, and, as hinted at in the paragraph about proliferation, cancer cells do not participate in this process. Instead, they continue to grow and divide and proliferate indefinitely, bypassing the programmed cell death and surviving to make the tumor larger and larger. If we could re-induce the programmed cell death, apoptosis, in the cancer cells, we could not only slow the growth of a tumor but even shrink the cancer until it is cured.

Knowing all of this, we can now search through the studies and research that have been performed concerning cannabinoids and cancer and make some sense of how helpful cannabis can been in treating cancer.

CHAPTER EIGHT: EFFECTS OF CANNABINOIDS ON CANCER PROLIFERATION

The amount and quality of research supporting the use of cannabinoids in fighting cancer could be surprising to some, though others have believed in the cancer-fighting and cancer-reducing properties of the substance for a while. Because there is so much research about the various ways in which cannabinoids can combat cancer, we will split the research into sections, including how it interrupts the four processes of cancer, management of cancer symptoms, and how it fights specific types of cancer.

This first section will be about cancer cell proliferation and how cannabinoids halt the process of cell growth and division in cancerous tumors. As mentioned previously, proliferation is the process of cancer cells growing and spreading.

General Studies On Preventing Cancer Proliferation

The active compounds in cannabis are anti-proliferative substances. This means that cannabinoids stop cancerous cells from growing and invading. A review[12] published in *Oncotarget* in 2014, scientists reported that CBD and THC, along with other cannabinoids present in the plant family, Cannabis, inhibit and prevent cancer cell proliferation. They specifically pointed out breast cancer, prostate cancer, and lung cancer as cancers that had been tested for the antiproliferative effects of cannabinoids, but they suspected that THC, CBD, and

---

[12] https://www.ncbi.nlm.nih.gov/pmc/articles/PMC4171598/

other types of cannabinoids would help with other cancer types as well.

Going back to previous years, in 2010, research[13] found that cannabis has an antiproliferative effect on the condition called endometriosis, which can often lead to cancer. Endometriosis is the formation of painful, precancerous lesions on female reproductive organs. Because cannabis reduced the proliferation of these deep-infiltrating lesions, it prevented cancer from cropping up in many women who were plagued by endometriosis.

Brain Cancer – THC And Glioblastoma

An important study[14] by Manuel Guzman and a team in Madrid, Spain, found that THC, or delta-9 tetrahydrocannabinol, reduced tumor proliferation in all nine of the patients it studied. Initiated because of some previous research findings by Cristina Sanchez, Guzman decided to study the antitumoral effects of THC on nine glioblastoma patients who failed to respond to standard brain cancer treatment of the time. He published the results in the British Journal of Pharmacology that *every* patient responded positively to some extent to the use of THC to reduce the proliferation of the brain tumors.

Lung Cancer - THC And Lung Cancer Proliferation

Not only does THC slow proliferation of brain cancer specifically, but Harvard scientists have also seen the

---

[13]https://www.ncbi.nlm.nih.gov/pmc/articles/PMC2993285/
[14]http://www.nature.com/bjc/journal/v95/n2/abs/6603236a.html

48

substance slow tumor growth in common lung cancer. According to Harvard researchers, THC "significantly reduces the ability of the cancer to spread." The research study[15] that they performed gave surprising results – THC targets unhealthy cells and leaves the healthy cells untouched and unscathed. This is in stark contrast to the highly toxic chemotherapy drugs that indiscriminately destroy brain, body, and cancerous tumor.

CBD And ID-1 Gene

In the Pacific Medical Center in San Francisco, California, Dr. Sean McAllister has been studying cannabinoids and cancer for ten years, trying to develop therapeutic interventions for various types of cancer. He has been backed by grants from the National Institute of Health and has a license from the US Drug Enforcement Agency, otherwise known as the DEA.

Using these privileges and financial resources, Dr. McAllister discovered that cannabidiol (CBD), the non-psychoactive component of cannabis, is a potential inhibitor of breast cancer cell proliferation and metastasis, as well as a potential means of halting tumor growth.

He went a step further and explained how this happens.[16] There exists a gene called the ID-1 gene that plays an essential role in cancer cell proliferation and metastasis. Initially, it is active during our

---

[15]http://www.nature.com/bjc/journal/v95/n2/abs/6603236a.html

[16] https://www.ncbi.nlm.nih.gov/pubmed/18025276

embryonic stage while we are in our mothers' wombs. However, it turns off when we are born and is supposed to stay switched off for the rest of our lives. In instances of cancer development, however, the ID-1 gene reactivates. When active during our post-partum state, that is, after we have exited the womb and the embryonic stage, the ID-1 gene causes cells to mutate, proliferate, and migrate to other parts of the body in a process of metastasis.

CBD *turns off* this gene. When CBD is introduced to a cell whose ID-1 gene is active, the gene suddenly stops acting and halts the progress of the cancer. According to Dr. McAllister, breast cancer is not the only cancer to exhibit this functioning of the ID-1 gene, but rather, "Dozens of aggressive cancers express this gene." Therefore, McAllister says, "Cannabidiol offers hope of a non-toxic therapy that could treat aggressive forms of cancer without any of the painful effects of chemotherapy."

Summary Of Cannabinoids And Proliferation

It seems that both THC and CBD are cannabinoids that can affect the proliferation of cancer. Research has discovered both to be effective in stopping the inhibition and prevention of the growth and further division of cancer cells which leads to tumor growth and to the cancer spreading. In terms of CBD, this seems to be connected to the shutting off of the ID-1 Gene. More research must be done to see what the particular effect of THC is upon cancer cells, but we can nonetheless say with relative certainty that cannabis and cannabinoids could be effective in stopping the proliferation of cancer.

CHAPTER NINE: EFFECTS OF
CANNABINOIDS ON METASTASIS

After looking at the effects of cannabinoids on proliferation of cancerous cells, you likely have high hopes for the effects of cannabinoids on metastasis. Indeed, the studies do not disappoint in terms of indicating the prevention of metastasis by cannabinoids. There is less research to this effect but it is nonetheless point to the possibility that cannabis could be used to prevent the spreading of cancerous tumors from one part of the body to another part of the human system.

One important thing to remember about metastasis is that a specific cancer spreading to another part of the body does not make the new tumor a cancer of that new part of the body. Breast cancer spreading to the lungs does not become lung cancer, and liver cancer spreading to the blood does not become leukemia. Each type of cancer develops separately. Therefore, it is not other types of cancer that we are trying to prevent. Rather, if we can prevent a cancer from spreading to the other parts of the body, we can halt the growth of the cancer from *affecting* other parts of the body. The goal is to minimize the amount of damage one type of cancer does to the body as a whole.

General studies[17] show that cannabinoids have anti-metastatic traits. Let us look a little deeper, though, into the research that has been done in this area of study.

---

[17] https://www.ncbi.nlm.nih.gov/pubmed/27070944

Spanish Research: Cannabinoids Preventing Metastasis

In 2012, researchers at Complutense University of Madrid in Spain discovered that cannabis effectively blocks metastasis of cancer. The results of the study[18] showed that the active cannabinoid components of cannabis blocked the cancer cells from breaking off and attaching to other parts of the body.

Leading author, Guillermo Velasco, also references not just one or two but *twelve* studies that have shown that cannabinoids block metastasis[19]. These studies are backed by each other's findings, showing that this is no fluke of research that shows that cannabinoids can be used in the blocking of metastasis of cancer.

How Cannabinoids Block Metastasis

In 2013, Italian scientists found[20] that CBD products particularly protect against cancer cells' "migration, adhesion, and invasion." Essentially, cannabidiol caused cancer cells that had broken off the main tumor to be unable to adhere to another part of the body and thereby prevented their invasion of other regions of the body. This reduces the risk of the cancer spreading and causing other organs to fail. Instead of focusing on many areas of cancerous tumor, doctors can focus on just the original tumor and effectively diminish it or remove it.

---

[18] https://www.ncbi.nlm.nih.gov/pubmed/22555283
[19] http://www.sciencedirect.com/science/article/pii/S0278584615001190
[20] https://www.ncbi.nlm.nih.gov/pmc/articles/PMC3579246/

Dr. McAllister's research also speaks to this, saying that by switching off the ID-1 gene, the CBD compound switches off the cancer cell's ability to grow and proliferate, which in turn makes it much harder for the tumor to break apart and spread.

Summary Of Cannabinoids And Metastasis

It seems that CBD wins the race of cannabinoids to prevent metastasis of cancer. CBD causes proliferation to halt and turns off the ID-1 gene such that it no longer expresses itself as cancerous cell growth. This reduction of the proliferation of cancer cells causes metastasis to be less of an issue, if it continues to be an issue at all. To summarize, cannabinoids and, in particular, CBD cause metastasis to slow or stop, which makes the cancer easier to contain and treat.

CHAPTER TEN: EFFECT OF CANNABINOIDS ON ANGIO-GENESIS OF CANCER CELLS

Angio-genesis is one of the main factors that allow cancer to grow and spread. Therefore, interrupting the process of creating blood vessels to the cancer is vital to stopping the growth of the tumor and even to disallowing the survival of cancer cells. Cannabinoids have been shown by studies to be effective against angio-genesis, just like they are able to combat cancer cell proliferation and metastasis.

Mini-Reviews in Medicinal Chemistry included this statement, in fact:

> [Cannabinoids] represent a new class of anticancer drugs that retard cancer growth, inhibit angio-genesis and the metastatic spreading of cancer cells...

It turns out that both THC and CBD are effective against angio-genesis. Let's look at how these two compounds affect the creation of new blood vessels to cancerous tumors and cells.

THC And Angio-Genesis

In 2008, a Spanish research team led by Cristina Blázquez discovered[21] that THC, the psychoactive component in Cannabis, diminished the ability of cancer cells to perform angio-genesis. In studying glioma cells, the cancerous cells of glioblastoma brain cancer, Blázquez and her team found that the cells had a harder time building blood vessels when affected by THC. Of course, this means that glioblastoma itself can be stopped or even diminished

---

[21]http://cancerres.aacrjournals.org/content/68/6/1945.long

through the use of THC, but the team went on to mention that the THC had that same effect on melanoma and carcinoma, two types of skin cancer.

CBD And Angio-Genesis

In 2011, researchers at Vanderbilt University found that CBD also stops angio-genesis in cancer cells.[22] It turns out that CBD interacts with the angio-genesis process of cancer cells in a different way than THC interacts with it, both effectively inhibiting the way that cancer cells build blood vessels to a tumor but each using different methods. This is an important discovery because it shows that cannabinoids have multiple methods of treating cancer, even between different cannabinoids.

Another study[23] by an Italian team and published in 2012 in the British Journal of Pharmacology shows that CBD uses multiple methods itself to fight angio-genesis. In essence, CBD discourages the expression of several genes that cause blood vessel growth in cancerous tumors. The study also found that the introduction of CBD caused cytostasis, which is the halting of cell growth, effectively ending the proliferation of cancer cells in a tumor or cancerous growth.

General Use Of Cannabinoids Against Angio-Genesis

A body of research[24] by a few scientists at the Medical College of the Polish Jagiellonian University, the

---

[22]https://www.ncbi.nlm.nih.gov/pmc/articles/PMC3366283/
[23] https://www.ncbi.nlm.nih.gov/pubmed/22624859
[24] https://www.ncbi.nlm.nih.gov/pubmed/28100841

oldest and most prestigious medical college in the nation, speaks to the general use of cannabinoids against angio-genesis and cancerous cell growth. The researchers concluded:

> Phytocannabinoids, endocannabinoids, synthetic cannabinoids and their analogues can lead to inhibition of the growth of many tumor types, exerting cytostatic and cytotoxic neoplastic effect on cells thereby negatively influencing neo-angiogenesis and the ability of cells to metastasize.

This means that various cannabinoids, including the ones found in the Cannabis sativa plant itself, ones found in the human body naturally, man-made cannabinoids, and any of their equivalents have antitumoral effects, causing the cancer cells to stop growing and even die because of the way they cut the cancer cells off from the blood source and keep them from travelling throughout the body.

Summary Of Cannabinoids And Angio-Genesis

The various cannabinoids seem to be mightily effective against angio-genesis. THC and CBD each have their own methods of stopping angio-genesis, and together they might effectively halt the development of blood vessels in cancerous tumors altogether. CBD, in fact, has multiple ways that it affects angio-genesis genes just by itself, cutting off the expression of the genes that develop blood vessels to tumors. According to the Polish researchers, the naturally occurring plant cannabinoids, synthetic cannabinoids, and the cannabinoids occurring within the human body all have potential to fight cancer and its angio-genesis.

CHAPTER ELEVEN: EFFECTS OF CANNABINOIDS AND APOPTOSIS

The natural process of apoptosis is necessary in the human body so that tumors do not arise, benign or malignant. When apoptosis is bypassed, tumors can form where functional cells would grow and cause a cancer to take over. Therefore, it is important to maintain apoptosis, or "programmed cell death," as part of the normal life cycle of cells in the human body. That is, causing the natural process of suicide in cancer cells can not only stop the growth of cancer but even shrink and eliminate cancerous tumors.

THC And Apoptosis Of Cancer Cells

Research from as far back as 1998 shows that THC causes apoptosis in cancer cells. When THC comes in contact with many types of cancer, it triggers the apoptotic response of the cells so that they literally kill themselves off.

The results that Cristina Sanchez and her team found at Complutense University in Madrid, Spain in 1998 were surprising. They were studying cell metabolism and the effect that THC has on the process. Instead of finding the results they were seeking, the team found that when glioma cells met THC molecules, they died. That is to say, the glioma cells, part of the aggressive brain cancer called glioblastoma, underwent cell death as a result of coming in contact with the delta-9 tetrahydrocannabinol. These results[25] in vitro, that is, in the lab, led to Manuel Guzman's later study in vivo, on live patients with glioblastoma, which we discussed

---

[25] http://scholar.qsensei.com/content/yo5d9

earlier and which resulted in all nine of the patients' cancer being affected positively.

How did this induction of apoptosis work? The researchers wrote their findings in a report[26] that essentially, THC connects to the cannabinoid receptor, CB1, in the cancer cells, causing apoptosis and breakdown of the cell material. The gene, SR141716, which usually causes cancer cells to forego their programmed cell death, was not effective in inhibiting the connection of THC to the CB1 receptor and therefore was no defense against the apoptosis-inducing chemical agent.

It has also been found at St. George's University in London, England, that THC, along with CBD, induces apoptosis in leukemia cells, or cancerous cells in the blood. This is an important discovery because, unlike other forms of cancer, leukemia does not form tumors and therefore is not affected by antitumoral traits of THC, CBD, and other cannabinoids.

Lastly, *Current Oncology* published an article in March, 2016 about a study[27] that showed that both THC and CBD compounds are effective against neuroblastoma, a common brain cancer in young children.

CBD And Apoptosis Of Cancer Cells

Current Oncology's article[28] about CBD and the induction of cancer cell death of neuroblastoma cells in children was a very important discovery, though

---

[26] https://www.ncbi.nlm.nih.gov/pubmed/9771884
[27] https://www.ncbi.nlm.nih.gov/pmc/articles/PMC4791143/
[28] https://www.ncbi.nlm.nih.gov/pmc/articles/PMC4791143/

not the first or only instance of CBD causing apoptosis in cancer cells. It was shown in this study that the apoptosis was induced by CBD both in vitro in the lab and in vivo in actual patients with neuroblastoma.

At the International Cannabinoid Research Society meeting in 2012 in Freiburg, Germany, about 300 research scientists gathered and gave their findings, discussing together the effects of cannabinoids on various conditions, including cancer. Italian researchers were reported to have said that CBD is "the most efficacious inducer of apoptosis" for prostate cancer. British researchers from Lancaster University agreed and added that CBD was also effective against colon cancer cells, inducing programmed cell death in them as well.

Another body of research[29] published in Chicago, Illinois, showed that CBD encourages apoptosis (programmed cell death), autophagy (the self-destruction, lit. "self-eating" of the cell), and the depression of the ID-1 gene.

Summary Of Cannabinoids And Apoptosis

According to the research that we have gathered through the years, it has become clear that THC, CBD, and other cannabinoids are able to combat cancer not only indirectly but directly by killing the cancer cells themselves through apoptosis. Using THC and CBD, not only can we inhibit proliferation of cancer cells, stop cancerous tumors from spreading to other parts of the body, and cut off the blood supply to cancer cells by disallowing the construction of blood vessels

[29]http://www.jpsmjournal.com/article/S0885-3924(13)00115-2/fulltext

to the tumor, but we can also attack the cancer cells directly and cause them to destroy themselves, shrinking the cancer and potentially curing it.

THC is effective against cancer cells in that it connects to the CB1 receptors, which are still responsive in the cancer cells, and thereby the THC induces the process of apoptosis. CBD has a similar antineoplastic effect, causing programmed cell death through autophagy, of the self-destruction of the cancer cells. It also inhibits regrowth through the suppressing of the ID-1 gene.

CHAPTER TWELVE: OTHER EFFECTS OF CANNABINOIDS AND CANCER

There are some other effects of cannabinoids that help manage cancer other than the interactions with cancer processes. Some of these ways are the boosting interaction of cannabinoids and traditional cancer-treating methods as well as the way that cannabinoids can treat cancer symptoms and treatment-induced side effects.

Interaction Of Cannabinoids And Traditional Cancer Treatments

Dr. McAllister, whom we met earlier in this book in discussing the discovery of the ID-1 gene's role in causing cancer, also did some research on the interaction of CBD and first line chemotherapy agents. He found that, while CBD is a powerful cancer treatment on its own, it is also powerful in boosting chemotherapeutic treatment such that it reduces the dosage necessary to treat cancer from a toxic to a manageable level. CBD interacts synergistically with traditional cancer treatments, therefore, to enhance the impact of the chemotherapy medications. CBD could be a powerful anticancer treatment itself, but its synergy with first line anticancer pharmaceuticals makes it even more precious a commodity and bestows on it even greater potential for usefulness in fighting cancer.

St. George's University in London did some experiments of a similar nature to Dr. McAllister's research and found that THC has a similar effect on traditional anticancer treatment options to CBD – it creates a deeper and more effective result from the chemotherapeutic medication. In particular, the team at St. George's tested THC as well as CBD alongside some leukemic medications, and they found that adding THC or CBD produced maximum effectiveness of the chemotherapy drugs on the leukemic cancer. This means that CBD and THC both work effectively on their own as well as synergize with other methods of cancer treatment, which could bring hope to the millions affected by and dying from cancer each year.

Cannabinoids And Cancer Symptoms

Cannabinoids can not only treat cancer but also alleviate the symptoms of cancer. According to the American Cancer Society,[30] the following effects of medical marijuana, that is, of Cannabis sativa and unprocessed phytocannabinoids, are accepted as true.

1. Marijuana can treat neuropathic pain, that is, pain from damaged nerves. Damage to the nerves takes place often in instance of cancerous tumors, as the tumor itself takes over and squeezes, pushes, and pressures the nerves that surround it.

2. Many patients who struggle with cancer also struggle with weight loss because they fail to eat enough to replenish their bodies' energy and fat and muscle supply. THC can help to alleviate this symptom, kicking up the appetite where once it was lacking. It does this by causing the metabolism to pick up very quickly, giving the user the "munchies." As you might

---

[30] https://www.cancer.org/treatment/treatments-and-side-effects/complementary-and-alternative-medicine/marijuana-and-cancer.html

guess, eating healthfully and eating enough are essential to surviving cancer, as the body cannot fight if it does not have the nutrients to do so.

Cannabinoids And Treatment Side Effects

The American Cancer Society also acknowledges some other uses for medical marijuana in terms of how cancer affects its victims. These uses are commonly to treat the side effects of chemotherapy, radiation, and other toxic treatments of cancer. These uses include the following effects:

1. Medical marijuana can be used to treat nausea and vomiting from chemotherapy and other toxic treatments for cancer. *Many* patients have reported using Cannabis sativa in its marijuana form to relieve feelings of nausea and habits of vomiting after radiation treatments.

2. Marijuana also helps treat the pain of various treatments so that patients who practice using marijuana do not need as many pain medications. Often the pain medications that are given to cancer patients are addictive, such as opioids, while marijuana is not shown to be largely addictive. Thus marijuana could be an excellent alternative to traditional pain medication in that, once the cancer is put into remission, the patient is not left with an addiction to pain medication.

Types Of Cannabinoids Legally In Use Today

The American Cancer Society as well as the great majority of doctors in America are quick to point out that the US Federal Government has outlawed the prescription, use, and possession of marijuana, which is defined as any substance containing the cannabinoid known as THC, or delta-9 tetrahydrocannabinol, or the lesser known delta-8 tetrahydrocannabinol. While the American Cancer Society supports the need for more scientific research on cannabinoids' use for cancer patients, it carefully words its promotion so as not to offend the Drug Enforcement Agency's sensibilities.

That being said, there exist two drugs legally in use today in the United States as well as one more under study that is being used in Canada and throughout some European countries that contain cannabinoids, albeit synthetic ones. These drugs are known as Marinol (Dronabinol), Cesamet (Nabilone), and Nabiximols.

Marinol, or *Dronabinol* is a gelatin capsule that is created from pharmaceutical THC. It is the only THC-containing substance that is legal according to the United States Government under DEA regulations. This drug is used to treat nausea and vomiting due to chemotherapeutic treatment. Being made from THC which is known for the "munchies," it also is able to treat weight loss and loss of appetite due to both cancer and the radiation treatment of cancer.

Cesamet, or *Nabilone* is a drug similar to THC that is taken by mouth. Like Marinol, it treats nausea and vomiting from chemotherapy. It is a synthetic drug, meaning it is lab-created and man-made. Because of this "pure" quality, the DEA does not see it as the same threat posed by marijuana. However, it also does not contain the cancer-fighting qualities of marijuana's natural cannabinoids.

Nabiximols is a cannabinoid drug that is under study in the United States, though it is used throughout Canada and in many countries in Europe. It is made of whole Cannabis sativa plant extract with about a one to one ratio of THC to CBD. It comes in the form of a spray that is supposed to cover the tongue and inside of the mouth. Nabiximols treats pain linked to cancer as well as muscle spasms and pain from multiple sclerosis.

The Verdict On Current Cannabinoid Medications

The DEA has severely restricted the use of cannabinoids, specifically THC. Even CBD products are not approved by the Food and Drug Administration for any pharmaceutical purpose. Therefore, doctors are unable to prescribe therapeutic doses of cannabinoids other than to minimize the effects of chemotherapy and radiation treatments, that is, to treat symptoms of nausea and vomiting. There is a lot of room for improvement federally.

At the state level, medical marijuana is legal in some states, along with recreational use of the substance in a smaller number of states like Colorado. Without the Drug Enforcement Agency's approval, however, the research to create pharmaceuticals from the plant, Cannabis sativa, or to find therapeutic uses for cannabinoids is very difficult to perform at any reasonable pace.

CHAPTER THIRTEEN: EFFECT OF CANNABINOIDS ON VARIOUS FORMS OF CANCER

Now we will take a look at some studies on various forms of cancer and how cannabinoids affect them individually. It has been implied that THC and CBD act differently on different types of cancer. Now we will begin to see what the differences are in the activities of cannabinoids in relation to different kinds of cancer cells.

Bladder Cancer

A study[31] in California was performed on over 84,000 men over an 11-year period. They reported a lower instance of bladder cancer among the cannabis users than among the general population. Calculating other factors out, they discovered that cannabis use makes a male 45% less likely to develop cancer of the bladder. Their conclusion was this:

> Although a cause and effect relationship has not been established, cannabis use may be inversely associated with bladder cancer risk in this population.

In other words, using marijuana on a regular basis could be associated with a lowered risk of developing bladder cancer for the male population.

Another study[32] showed results that CBD has a positive effect on bladder cancer, causing apoptosis of the urothelial carcinoma cells, that is the cancerous bladder cells. The transient receptor potential

---

[31]https://www.ncbi.nlm.nih.gov/pubmed/25623697?dopt=Abstract
[32] https://www.ncbi.nlm.nih.gov/pubmed/20546877

vanilloid 2 (TRPV2) in the T24 cancer cells causes proteins that are vital to the cell's survival and growth to be channeled into the cell. Through the introduction of CBD to the T24 cells, calcium flows into the T24 cells, causing a decrease in the TRPV2 mRNA (a type of DNA) expression. This decreased the viability of the T24 cells, that is, the ability of the cells to live and function. In fact, through a continued influx of calcium into the cell because of CBD, T24 cells underwent apoptosis.

This means that not only could CBD stop the proliferation of T24 cells by disallowing them to receive the proteins they need to divide, but CBD could also work towards curing bladder cancer through causing the programmed cell death of the T24 urothelial carcinoma cells.

Brain Cancer

There are many studies concerning brain cancer (glioblastoma) and cannabinoids. Some have been mentioned already, but here are some more studies and their results for your perusal. Together, they show that brain cancer could be effectively treated using THC and CBD.

First, we will investigate the effects of CBD on glioma cells, as we have already seen from Cristina Sanchez's research earlier in this book that THC is an effective compound in causing apoptosis of glioma cells. Relating to CBD and glioma, in one study[33] in 2003, an Italian team discovered that CBD's anti-proliferative effect on glioma cells was due to the way it induced apoptosis in the cancerous cells. The CB2

---

[33] https://www.ncbi.nlm.nih.gov/pubmed/14617682

receptor antagonist, the protein that works against CBD adhering to the CB2 receptor, mitigated some of this effect, but the CB1 receptor antagonists were no match for the CBD.

A study in 2005[34] concerning cannabidiol and glioma cells found another effect of CBD on glioblastoma. While CBD works with the CB1 and CB2 receptors to prevent survival of the glioma cells by inducing apoptosis, CBD works in a different way to control migratory spread of the cancer. Instead, the anti-metastatic effects of CBD seemed to be independent of the traditional CB1 and CB2 receptor interaction, meaning that CBD has more than one way of interacting with glioma cells and fighting cancer.

A 2013 study[35] showed that CBD interacted in yet another way with glioma cells, specifically the kinds called U87-MG and T98G. Instead of simply causing apoptosis outright, the introduction of the CBD compound stopped glioma cell growth, invasion, and angio-genesis. The researchers became excited at this discovery because it proved, in their words, "new insights into the antitumor action of CBD, showing that this cannabinoid affects multiple tumoral features and molecular pathways." This is to say, CBD interacts in a variety of ways to combat cancer.

Another study in 2013[36] gave results that administering CBD along with traditional pharmaceuticals for treating cancer can increase the effectiveness of the original treatment. As with bladder cancer, introducing calcium to the transient

---

[34] https://www.ncbi.nlm.nih.gov/pubmed/15700028
[35] https://www.ncbi.nlm.nih.gov/pubmed/24204703
[36] https://www.ncbi.nlm.nih.gov/pubmed/23079154

receptor potential vanilloid 2 (TRPV2) can be very effective. Such a process allows more of the cell-killing substance to enter the cell and thus causes cell death, or apoptosis, while leaving the healthy cells, the astrocytes, unscathed. The conclusion of this research study was that using CBD alongside cytotoxic pharmaceuticals increased the uptake of the cytotoxic chemicals into the cancer cells and also potentiated, or enhanced, the substance's ability to cause apoptosis.

A Canadian study in 2013[37], in reference to the ID-1 gene, showed that knocking out the ID-1 gene severely limits the glioblastoma's ability to proliferate. The research showed that cannabidiol, CBD, causes the knockdown of the ID-1 gene and therefore results in the limitation of the invasive and self-renewing properties of the cancerous cells.

A very recent study from 2016, published in 2017[38], showed a few things, but most importantly, it again showed that including CBD alongside traditional pharmaceuticals can be very effective. It demonstrated that introducing cannabidiol alongside DNA-damaging agents (a traditional treatment of cancer), created a synergy of cell-killing and anti-proliferating effects. The concentrations needed to be specific, but when in the right range, the CBD caused the DNA-damaging agent to be more effective than it would be on its own.

One study[39] builds on the fact that delta-9 tetrahydrocannabinol (THC) is effective against

[37] https://www.ncbi.nlm.nih.gov/pubmed/23243024
[38] https://www.ncbi.nlm.nih.gov/pubmed/27821713
[39] https://www.ncbi.nlm.nih.gov/pubmed/20053780

glioblastoma. The researchers heard conflicting data about whether other cannabinoids inhibited or enhanced the effects of THC on glioma cells. Therefore, they performed a test using cannabidiol and THC on glioma cells. They found that CBD enhanced the ability of THC to fight glioma cell proliferation and survival. While THC by itself performed certain tasks in relation to the CB1 and CB2 receptors and thereby inhibited glioma cells' ability to divide and proliferate, CBD modulated THC's actions so that together the two compounds had a different effect than each one individually. Together, they caused apoptosis of the glioma cells, leading toward the curing of the cancerous tumor.

Other studies show that both THC and CBD are effective against glioma cell division and the growth of the brain cancer tumor, glioblastoma. Another 2013 study[40] showed that local injection of THC, CBD, or a one-to-one ratio of the two substances reduced glioma tumor growth five times as efficiently as oral consumption.

Breast Cancer

Breast cancer is one of the most prominent cancers in the modern day. Whether because breast cancer is more publicized or because it is more widespread, the fact remains that a large amount of research has been geared toward finding a cure for breast cancer.

A study[41] was reported by the Institute of Biomolecular Chemistry to the Italian National Research Council in 2006 that tested the efficiency of

[40] https://www.ncbi.nlm.nih.gov/pubmed/23349970
[41] https://www.ncbi.nlm.nih.gov/pubmed/16728591

five different cannabinoids against breast cancer. Cannabidiol (CBD) was the frontrunner of the experiment, causing apoptosis of the breast cancer cells, otherwise known as breast carcinoma. Behind CBD came cannabigerol (CBG) and cannabichromene (CBC), in that order. CBD's ability to kill breast carcinoma cells while minimally affecting the surrounding healthy cells could give great hope to breast cancer patients in the future, minimizing the damage to surrounding tissues while killing off the breast cancer itself.

In 2007, there was a study[42] performed in California by Dr. McAllister that found that CBD was anti-proliferative and anti-metastatic towards breast cancer cells in vivo and in vitro. He was excited at this discovery because, in his words, "CBD represents the first nontoxic exogenous agent that can significantly decrease Id-1 expression in metastatic breast cancer cells leading to the down-regulation of tumor aggressiveness." That is, unlike chemotherapy or radiation, CBD is non-toxic and therefore poses much less risk to the patient than traditional cancer treatments.

In 2010, Dr. McAllister performed another research study[43] to discover the means by which CBD reached its anti-proliferative and anti-metastatic effects. The research was performed in vivo rather than in a petri dish (in vitro), and it found that, "CBD inhibits human breast cancer cell proliferation and invasion through differential modulation of the extracellular signal-regulated kinase (ERK) and reactive oxygen species (ROS) pathways." This then causes the decrease in ID-

---

[42] https://www.ncbi.nlm.nih.gov/pubmed/18025276
[43] https://www.ncbi.nlm.nih.gov/pubmed/20859676

1 gene expression, as well as the emphasis on the ID-2 cell differentiation gene expression which causes the cells to form according to their function as mammary cells rather than as malignant cancer cells. In this way, CBD uses two methods of downplaying the ID-1 gene – through interactions with the extracellular signal-regulated kinase and through the effects it has on the reactive oxygen species. It also has two different methods of preventing proliferation and metastasis – through the knockdown of the ID-1 gene in cancer cells and the promotion of the ID-2 gene expression toward cell differentiation.

In 2012, in the Department of Molecular Biology of Daiichi University in Japan, an interesting discovery[44] was made concerning cannabidiolic acid, otherwise abbreviated as CBDA. The researches at Daiichi University concluded:

> The data presented in this report suggest for the first time that as an active component in the cannabis plant, CBDA offers potential therapeutic modality in the abrogation of cancer cell migration, including aggressive breast cancers.

This is exciting news because it means that unaltered Cannabis sativa plants could be used to treat breast cancer. Rather than having to smoke or otherwise inhale the plant, or create expensive oils out of the plant for injection, doctors can administer the cannabidiolic acid to the patient directly and see results of antiproliferation.

For another study published in 2016[45], Daiichi University researchers cooperated with two other

---

[44] https://www.ncbi.nlm.nih.gov/pubmed/22963825
[45] https://www.ncbi.nlm.nih.gov/pubmed/27530354

institutes of higher learning to continue their research of the utility of CBDA in treating breast cancer. The scientists identified cyclooxygenase-2 (COX-2) regulation as a key factor in cannabidiolic acid's influence over the cancerous cells in breast cancer. According to their findings, COX-2 expression has been found in 40% or more of breast cancer cases, and CBDA down-regulates and chemically inhibits its expression. This might be how CBDA causes an antiproliferative effect on breast cancer.

CHAPTER FOURTEEN: METHODS OF CONSUMPTION

There exists a number of methods of consumption that cause CBD oils and other cannabinoids to enter the human biological system. Each method engages the biology of the human system in a unique way, interacting with the endocannabinoid system, the liver, and other organs. The effects of the various methods of consumption. For example, mood and motor skills are affected by the eating, smoking, and inhalation of cannabinoids, as well as the senses of pain and pleasure. Appetite and metabolism function differently due to the consumption of cannabinoids as well. CBD and THC also affect the reproductive system and immune system. Sleep cycles and sleep quality could also improve as a result of the ingestion or inhalation of THC, CBD, and other cannabinoids.

Once in the system THC, CBD, and the other cannabinoids found in Cannabis sativa and medical marijuana function as neurotransmitters. There exists an endocannabinoid called anandamide that is native to the human body that stimulates the CB1 and CB2 receptors in human cells. THC, CBD, and other cannabinoids mimic this compound, anandamide, stimulating the CB1, CB2, and other cannabinoid receptors in various ways, causing varied results. Specifically, THC stimulates the brain to produce serotonin. THC also causes the creation of dopamine, which combats negativity by affecting an individual's emotions and behavior. CBD also interacts with the special cannabinoid cell receptors like CB1 and CB2 and thereby causes various effects on the central nervous system, though without being psychoactive like its companion, THC.

The various methods of consumption that we will investigate are the following: inhalation methods including smoking and vaporization, and ingestion methods which include consuming edibles, consuming ingestible oils and tinctures, and consuming raw cannabis.

Smoking – An Inhalation Method

Smoking can be performed through a variety of means: through bongs, through hookahs, through pipes, through rolled joints, and through blunts, for example. This is the traditional method of recreational cannabis consumption, though an individual might struggle to reach therapeutic doses using this method. Smoking activates the THC in the cannabis plant material through heating the THCA (delta9-tetrahydrocannabidolic acid). The flame from the burning plant material heats the THCA structure to convert it to active THC. This burns much of the THC and CBD, though resulting in less being available for use by the human biological system.

There are a few drawbacks to this method. First, as mentioned, much of the THC produced by this method is non-bioavailable, meaning that a good amount of the resulting THC is not usable by the human body. As a result, individuals using Cannabis sativa to treat medical symptoms and conditions like cancer will have a hard time reaching therapeutic levels of THC, of CBD, and of other cannabinoids.

In addition, this method exposes the mouth, lungs, and other tissues directly to the smoke produced by the burning of the cannabis plant material. The cannabinoid receptors themselves in these regions of the body, both CB1 and CB2, are also exposed to the smoke. No studies have connected the smoking of cannabis to lung cancer, and in fact the use of cannabis has been shown to reduce the effects of lung cancer as we have seen. However, there is certainly evidence that microscopic damage is done to the lungs when a person smokes cannabis. These damages start to heal itself as soon as one stops smoking the substance, but it is damaging to the lungs when smoked on a consistent or constant basis.

Finally, because of the damage to the lungs or as an independent symptom of the smoke coming in contact with the body tissues, smoking cannabis causes inflammation and increased mucus production. It also causes bronchitis-like symptoms, though it is not the same as having an infection of bronchitis itself. That is, there is no bacteria causing the bronchitis symptoms, but the coughing and soreness associated with bronchitis exist with smoking of cannabis nonetheless.

The American Cancer Society has admitted some benefits of smoking cannabis associated with the disease and treatment of cancer, however. It is acknowledged by the Society that smoking marijuana can treat the nausea that results from chemotherapy and limit the amount of vomiting a person experiences as a result of such radiative methods of treatment. In addition, inhaled marijuana, whether smoked or vaporized, can treat neuropathic pain, the pain from damaged nerves. People who smoke marijuana tend to need fewer pain medications and become addicted less often to opioids and other medications that alleviate pain symptoms. The

American Cancer Society also acknowledges that smoking marijuana can cause an increase in appetite (due to the THC content) and thereby treat the weight loss that often accompanies cancer.

To summarize, smoking cannabis is an inefficient means of putting THC, CBD, and other cannabinoids into the bloodstream. Additionally, it is potentially harmful to the lungs and to other tissues that interact with the smoke, including the CB1 and Cb2 receptors. However, it does treat some of the superficial symptoms of cancer, alleviating pain, increasing appetite, and limiting the amount of nausea and vomiting that a person experiences with chemotherapy. While smoking marijuana might not treat the cancer itself through anti-proliferation, anti-angio-genesis, anti-metastasis, or promotion of apoptosis, it can certainly help relieve the symptoms of other treatments like radiation and chemotherapy as well as diminish the need for addictive pain medications.

Vaporization – An Inhalation Method

Inhaling THC, CBD and other cannabinoids through vaporization might be a good alternative if you are looking to avoid the effects of smoke on your lungs and body. By engaging in this method of inhalation, you can reduce or eliminate the potential risks associated with smoking while still enjoying the inhalation method of consuming cannabis. A vaporizer heats the cannabis plant material to somewhere between 315 and 410 degrees Fahrenheit until the active ingredients, that is, THC, CBD, and other cannabinoids, evaporate. This causes the THCA and CBDA, the acid forms of THC and CBD, to convert to their active forms. This also helps reduce the amount of wasted cannabinoids, as some vaporizers are known to produce as much as 95% THC.

The resin of the cannabis is heated in such a way that the plant material containing the cannabinoids does not burn, so you do not have to worry about the effects of smoke on your body. This eliminates the effects of potentially carcinogenic ash, which is a counterproductive substance, and it reduces the number of free radicals that enter an individual's system, since many of the free radicals are derived from smoke. This slows damage to cells as well as disengages from the aging process that is sped up by smoking.

Once evaporated into the air, the steam that is produced by the vaporizer is inhaled by the individual user. Then, the THC, CBD, and other cannabinoids enter the bloodstream through the lungs, making their way to the brain quickly and engaging the central nervous system. A person might still struggle to reach a therapeutic dose of cannabinoids using the vaporization method however, as the bioavailability of the cannabinoids is still comparatively low. More efficient than the other inhalation method, smoking the cannabis plant material, still only about 30% of the THC inhaled through the steam of the vaporizer is bioavailable, or able to be used by the human biological system.

The American Cancer Society acknowledges the usefulness of vaporization (and smoking) of marijuana in reducing neuropathic pain that comes from the damaging of nerves through cancer itself, radiation, and chemotherapy techniques. Like smoking, vaporization does not treat the cancer's proliferation, metastasis, angio-genesis, or cause apoptosis, but it can help the process of cancer treatment be less painful without making you turn to opioids and other addictive drugs.

To summarize, vaporization is a safer method of inhalation than smoking, avoiding the possibly carcinogenic effects of smoke while maintaining the pleasurable experience cannabis inhalation techniques. You still are able to enjoy the pleasant aroma of cannabis through this method, but it might not be the best method if you are hoping to create a medicinal level of cannabinoids in your system. The bioavailability of the THC, CBD, and other cannabinoids is limited, so you might want to turn to more concentrated methods to find therapeutic levels of cannabinoids treatment of the cancer itself rather than just diminishing the effects of the cancer.

Edibles – An Ingestion Method

Edibles are any food item that contains cannabis and that can therefore transport cannabinoids from the plant material to the human biological system after heating. Consuming edibles is one of the most potent methods of delivery of THC, CBD, and other cannabinoids in existence, but it is still limited. One thing that *must* be done before consuming an edible is it must be heated by such a means as being baked into a cake or other food. If it is not baked or cooked, then it would fall under the "raw cannabis" category, which we will discuss later in the chapter. The process of heating causes the decarboxylating process, which transforms THCA, the acidic form of delta-9

81

tetrahydrocannabinol, into THC, the active cannabinoid. Other acidic cannabinoids like CBDA and CBCA are also converted to their active forms, CBD and CBC, through this process.

With this method, THC, CBD, and other cannabinoids are metabolized by the liver. As a result, more of the cannabinoids are converted to usable form by the individual's body, and therefore more of the compounds are bioavailable. Additionally, the liver converts more of the THC to the metabolite called 11-hydroxy-THC, which is four to five times more psychoactive than traditional THC.

Bioavailability is still limited to between 4% and 12%, meaning that the great majority of the compound is not available to the body for processing. In other words, THC becomes a very potent metabolite compound, but the medicinal and therapeutic effects of cannabis on cancerous cells are limited with the consumption of edibles method of consuming cannabinoids.

In addition, another drawback to consuming cannabis through edibles is that they must process through your liver. This process takes much longer than the process that happens in smoking and vaporization. With inhalation methods, the THC, CBD, CBC, and other cannabinoids are received by the lungs and transferred directly to the bloodstream without being processed, allowing them to reach the central nervous system rather quickly. With ingestible consumption, the liver converts THC to 11-hydroxy-THC, the powerful metabolite, but it takes time to do so and therefore the active cannabinoids take longer to reach the blood and central nervous system.

To put it plainly, edibles are a wonderful method for recreational use of cannabis, but again you will struggle to consume an appropriately high level of THC, CBD, and other cannabinoids using this method. You might very well receive the high from eating cannabis in edibles because of the potent 11-hydroxy-THC metabolite that it produced, but you will not receive the therapeutic benefit of treating cancer cells as easily.

Ingestible Oils and Tinctures – An Ingestion Method

Ingestible oils and tinctures[46] are the generally the most accepted forms of medical cannabis. They are highly concentrated extractions of THC, CBD, and other cannabinoids from the cannabis plant. Often, to create this type of substance, alcohol is used as a solvent to create an ultra-high concentration dose of cannabinoid. The alcohol separates the essential oils, such as CBD oil, from the plant material to create a product with a very high number of cannabinoids.

It is best to ingest cannabinoids with lipids or fats, which are present in the resin from which the cannabinoid oil is separated. Therefore, some methods leave a bit of the resin in with the cannabinoid oil. Another method is to infuse the cannabis extract with a carrier oil. Such carrier oils include coconut oil or an MCT, a medium-chain-triglyceride. Cannabinoids are fat soluble rather than water soluble, so encasing the cannabinoids in fat allows for faster delivery of the cannabinoids to the nervous system and biological system in general. It

---

[46] http://amzn.to/2vALFyd

also improves bioavailability. The liposomal oils are possibly the most potent method of delivering THC, CBD, and other cannabinoids to the human body.

Ingestible oils come in capsules[47] or as oils[48] that can be placed on the skin. They are known to result in therapeutic levels of cannabinoids in the system, which is good news for cancer patients. This is your best bet in terms of fighting cancer or other conditions, because it works efficiently. It is the most commonly used form of consumption for medical purposes as a result. If you are looking to use cannabis medically or therapeutically, you will likely be using ingestible oils or tinctures. Recreationally this would not be your choice, as these oils tend to be high in CBD, the non-psychoactive component of cannabis, which counteracts any THC that is contained in the oil.

Raw Cannabis – An Ingestion Method

Ingesting raw cannabis without heating it preserves the cannabinoid acids. Before it is heated, THC is delta9-tetrahydrocannibidolic acid, also known as THCA. Other cannabinoids, like CBD and CBC are found in acid form as well as cannabidiol acid (CBDA), for example. Only heat and age can convert these acids to their "active" forms.

Consuming raw cannabis puts hundreds of time more cannabinoids in the biological system than heating the herb through smoking, vaporizing, or baking edibles. Raw acids also act differently in the body that the active cannabinoids, THC, CBD, CBC and others. The

---

[47] http://amzn.to/2toxVtL
[48] http://amzn.to/2tdZ9PF

acids engage the endocannabinoid system in a different way than active cannabinoids, as their interaction with the cannabinoid receptors, CB1 and CB2 is unique to cannabinoid acids. For example, THCA is shown to have no psychoactive element, though THC has a definite psychoactive feature.

Most of the time, the way that raw cannabis is consumed is through juices and smoothies: they are essentially unbaked, unheated edibles. More studies and research are needed to understand the exact effects of raw cannabis on the human body and nervous system. It is unclear whether this method of consumption is effective against cancer and other conditions, though it might be a pleasurable experience nonetheless. This is not the recommended route for treating cancer, because most of the studies that have been performed have all used active cannabinoids like THC and CBD, so only those substances, rather than the substances *and* their acids, are known for anti-proliferative, anti-metastatic, anti-angio-genetic, and pro-apoptotic effects. However, you might recall that some studies showed that CBDA can be an effective substance in fighting breast cancer, among other uses it might have.

The Verdict

Ingestible oil and tinctures come out as the clear winner for treating cancer. They are the most potent form of cannabinoid as they are extremely concentrated, and they do not have the potentially counterproductive, carcinogenic or otherwise damaging qualities of smoking the substance of cannabis. Additionally, ingestible oil and tinctures are comparatively fast-acting ingestibles because, as fat-soluble liposomal oils encased in fats, they are quickly absorbed into the liver. They are also very efficient, as the liver converts the THC to 11-hydroxy-THC, which is extremely potent. The cannabinoids are then sent into the bloodstream and, from there, make their way to the brain and into the endocannabinoid system. This can treat a generalized cancer.

If the cancer has not spread very far, injections of CBD and THC are known to take place such that the consumption of THC and CBD is localized. This reduces the potential side effects of THC, causing it to be less psychoactive. All around, oils and concentrated injections of THC, CBD, and other cannabinoids is the clear choice for cancer patients seeking to fight the proliferation, metastasis, angio-genesis, and skipping of the apoptotic stage of cell life in cancer cells.

CHAPTER FIFTEEN: BENEFICIAL STRAINS

Having looked at the various methods of consumption, it would be beneficial to now look at the various strains of Cannabis that might be helpful to cancer patients. We will split this chapter into sections of the symptoms that cancer patients face because of their treatment regimen and because of the cancer itself. We will look at Cannabis strains for the general treatment of cancer and chemotherapy side effects, strains to treat pain, strains to help with nausea, strains that treat appetite and weight loss, strains that help with depression, and strains that can ease symptoms of fatigue.

Overall

Chemo is the winning strain when it comes to treating the side effects of cancer. It is also called UBC Chemo, referring to the University of British Columbia where is it said to have been developed in the 1970s specifically to treat chemotherapy side effects. It has extremely high THC levels, between 18 to 21%. Chemo treats nausea, appetite loss, insomnia, and pain. Its effects are ones of relaxation, sleepiness, hunger, happiness, and euphoria, thereby treating anxiety, insomnia, appetite loss, and depression as well as pain. This all-around winner is an excellent choice if you are going through chemotherapy as your cancer treatment regimen.

Pain

Harlequin is a strain that is high in CBD, with a 5 to 2 ratio of CBD to THC. It is Cannabis sativa dominant with some Cannabis indica mixed in. It provides clearheadedness and alertness, creating a relaxing effect without sedating the user. The key effects of Harlequin are, therefore, relaxation, happiness, focus, upliftedness, and energy. It is highly able to treat pain and stress, and it can also help with depression, inflammation, and fatigue. The way it treats inflammation also helps to reduce pain, and the manner in which it treats depression helps reduce the side effect of low mood that often comes with pain.

ACDC is a Cannabis sativa dominant strain that contains very high levels of CBD. CBD has been measured in levels up to 19% in this strain. The ratio of CBD to THC is around 20 to 1. Because it is high in CBD, it can treat pain and inflammation without the side effect of intoxication due to the psychoactive effects of THC. CBD helps to counteract the small amount of THC high that the use would experience from this strain. The effects of this strain are relaxation, upliftedness, happiness, focus, and energy, and the strain, ACDC, is best at treating stress and pain. It is therefore ideal for pain management. It can also treat inflammation, depression and headaches as well.

Blackberry Kush is a Cannabis indica dominant strain, a mixture of the Afghani and Blackberry strains. The indica has strong effects on pain, reducing and even eliminating it. The effects of this strain include relaxation, sleepiness, happiness, euphoria, and hunger. The strain, Blackberry Kush, is therefore good at treating stress, insomnia, pain, depression and lack of appetite. This would be an ideal strain if you are suffering from pain as well as appetite loss, depression and insomnia.

God's Gift is a strain that came from Granddaddy Purple and OG Kush and was developed in California. It is full of THC, with a THC content rating of 18 to 22%. Its effects are relaxation, happiness, sleepiness, euphoria and a bit of upliftedness. It is an ideal strain for treating stress and pain, both of which come in abundance with cancer and cancer treatment. It can also treat insomnia with great gusto, as well as depression and nausea, which can also be side effects of cancer and cancer treatment.

Nausea

Northern Lights is a great treatment for nausea. It is pure indica with good amounts of THC, resulting in a bit of psychoactive effects. It also has effects of relaxation, sleepiness, happiness, euphoria, and upliftedness. It relaxes the muscles in the body through its psychoactive effects and therefore is ideal for treating nausea. It also is good for treating stress, pain, depression, insomnia, and lack of appetite as well.

Blueberry Diesel is another strain that can treat nausea. It is a cross between Blueberry and Sour Diesel. It is uplifting, creates feelings of happiness, relaxes the user, induces euphoria, and increases creative thinking in the user. It is ideal for treating nausea and depression as well as for helping with fatigue, seizures, stress, and pain. This is an excellent choice if you are feeling nauseous from cancer treatment as well as suffering from depression, either independently or as a result of the cancer treatment process.

Super Lemon Haze is a strain that uplifts and gives a psychoactive effect with its high THC content. It is a mixture of Lemon Skunk and Super Silver Haze, and is Cannabis sativa dominant but have some Cannabis indica in it as well. Its effects include happiness, improved energy, upliftedness, euphoria, and increased creativity. As a result, Super Lemon Haze is extremely effective for treating depression and stress, as well as for treating pain, fatigue, and lack of appetite. It is also very effective against nausea, which is why it is included in this section.

Jillybean is a strain that resulted from the cross between Orange Velvet and Space Queen. It helps clear the mind and reduces or eliminates nausea. Its effects are ones of happiness, euphoria, and upliftedness as well as some energy improvement and creativity mixed in. It is extremely effective in reducing stress on the body, which is a major source of nausea. Jillybean also reduces depression, pain, and fatigue, as well as going for the source of nausea itself.

Loss of Appetite

Bubba Kush is of Afghani descent, but no one really knows its origins. It supposedly came from Northern Lights in some capacity, but its true origins are ambiguous. It is a tranquilizing strain with high THC content, and it is a Cannabis indica strain. The effects of this strain are of relaxation, happiness, sleepiness, euphoria, and hunger. Therefore, the strain is extremely effective at treating pain, stress, and insomnia, and it is also great for depression and loss of appetite as well as weight loss due to cancer and cancer treatments. It will give you the "munchies" and help you to consume the calories that you need to fight the cancer and keep on weight.

Skywalker OG Kush is a Cannabis indica dominant strain that is great for giving you the munchies and helping you to eat even when you do not feel up to it. The strain's effects include relaxation, happiness, euphoria, upliftedness, and tingles. Besides helping you with appetite loss, you can look to Skywalker OG Kush for stress-relief, relief from depression, pain relief, reduced or eliminated insomnia, and headache relief.

Granddaddy Purple is a strain that we have mentioned already as a parent strain, but it is also helpful in and of itself in helping create appetite in the user. Granddaddy Purple is a mixture of Purple Urkle and Big Bud. It is high in Cannabis indica, so it is excellent for treating pain and loss of appetite. Its effects are relaxation, sleepiness, happiness, euphoria, and hunger. Therefore, it is an excellent source for stress relief, insomnia reduction, pain relief, relief from symptoms of depression, and, of course, regaining of appetite.

Depression

Super Silver Haze is another Cannabis strain that we have already mentioned as a parent strain but that has great properties for cancer relief on its own. Specifically, it is great for reducing, relieving, and eliminating symptoms of depression. It creates happiness, euphoria, upliftedness, energy improvement, and creativity in the user. Therefore, it is great for stress relief, depression, fatigue, pain, and loss of appetite.

Chernobyl is another strain of Cannabis that can be excellent for relieving your symptoms of depression. It was originally bred by TGA Genetics, and its effects are happiness, euphoria, relaxation, upliftedness, and energy improvement. Therefore, it is ideal for treating stress, pain, and depression as well as for helping with headaches and loss of appetite.

Pennywise is high in CBD and is a Cannabis indica dominant strain. It contains about a 1 to 1 ratio of CBD to THC. It creates feelings of relaxation, happiness, upliftedness, sleepiness, and euphoria, and it is ideal for treating pain, stress, and depression. It can also be used for helping with inflammation and insomnia.

Fatigue

Strawberry Cough is an excellent, Cannabis sativa dominant strain for fighting fatigue. It creates feeling of upliftedness, happiness, euphoria, energy, and relaxation. Therefore, it us ideal to help with stress and depression, both of which often result in fatigue. It can also help with pain and lack of appetite.

Pineapple Express is one of those strains that many people have heard of even if they have not delved into the world of Cannabis as a result of the movie by the same name. Pineapple Express is not entirely what it was portrayed to be in the movie, but instead its effects are ones of happiness, upliftedness, euphoria, relaxation, and energy. Therefore, this strain is great for stress and depression, like the strain, Strawberry Cough, as well as for directly treating fatigue, pain, and loss of appetite.

Chocolope is the final strain we will discuss. It is excellent and even ideal for dealing with fatigue caused by cancer and cancer treatments. It creates feelings of happiness, upliftedness, energy, creativity and euphoria. Because of these effects, it is great for treating depression and stress as well as for helping with fatigue, pain, and nausea.

There are many other strains out there as well that might help with your cancer symptoms, so some experimentation is necessary to see what works best for you. These are a great starting point, though, and should help you discover how to best treat your chemotherapy and cancer side effects.

SUMMARY

A vast array of topics has been covered in this book, so to put it all together, we might need to review where we have been and what we have learned.

In Chapter One, we looked at what is CBD oil. We found that it is a cannabinoid and is a cousin of THC, but it does not have the psychoactive element of the other compound. We discovered that CBD is found in Cannabis sativa and other cannabis plants.

In Chapter Two, we learned about what cannabinoids are, and we found that the human body contains a system, called the endocannabinoid system, that responds specifically to cannabinoids like anandamide and externally occurring phytocannbinoids. We found that Cannabis sativa and other cannabis plants contain such phytocannabinoids as THC, CBD, CBC, and other cannabinoids, and we took a look at some varieties of active cannabinoids in the Cannabis sativa plant as well as their effects on the human body.

After that, in Chapter Three, we delved into a short bit of history of Cannabis sativa and dispelled some misconceptions. For example, we discovered that there is not a dichotomy between THC and CBD, such that THC is the "bad," dangerous drug it is made out to be while CBD is the "healthy" therapeutic substance that we should all be using in our diets. Instead, we found that THC has just as many therapeutic uses as CBD, and in fact, THC and CBD work best together.

In Chapter Four, we found out how CBD oil is made. We learned about various methods of extraction of the cannabinoid oils and found that some methods have greater purity and fewer impurities than other methods. We looked at how some have drawbacks that are not present with other extraction methods.

In Chapter Five, we looked at how cannabinoids work. We found out how they interact with the brain and nervous system and how they create the effects that they do. The CB1 and CB2 receptors seem to play a vital role in these interactions.

In Chapter Six, we gained some insight into how to choose the best oil for your needs, considering purity and price as well as some other factors.

In Chapter Seven, we started the discussion of cancer. We looked at the mechanisms by which cancer grows and spreads, learning about proto-oncogenes, tumor repair genes, and DNA repair genes. We found that when these genes mutate, cancer can occur. We also discovered the four processes that are essential to cancer's survival and growth – proliferation (or division of the individual cancer cells to form new cells), metastasis (the breaking off of part of the tumor and migration of the cancer to another part of the body), angio-genesis (the development of blood vessels to the tumor in carry in the necessary proteins for growth and survival and to remove waste), and the lack of apoptosis (programmed cell death) in the lifecycle of the cancer cells.

Chapter Eight brought us through a discussion of cannabinoids and their antiproliferative properties. We found numerous studies that allowed us to conclude that both CBD and THC could be very active and efficient anti-proliferative substances when they come in contact with cancer.

In Chapter Nine, we turned to a discussion of cannabinoids and their anti-metastatic properties. Again, numerous studies helped us realize that cannabinoids including THC and CBD could be very effective in keeping cancers from metastasizing and thereby taking over other parts of the body. Containing the original cancer at its origin could be very helpful to doctors who are trying to treat cancer patients.

Chapter Ten led us through some information about the anti-angio-genesis effect that THC and CBD have on cancer. We found that this is important because cutting off the cancer from the blood supply, or at least preventing it from building new blood supply, keeps it from growing and can even induce cell death. We discovered too that THC and CBD have different methods of stopping angio-genesis, which is even more exciting.

We took a look at cannabinoids and apoptosis, programmed cell death, in Chapter Eleven. We found even more studies that showed that THC and CBD can be effective in inducing apoptosis in otherwise endlessly proliferating cancer cells. Studies on brain cancer cells called glioma cells were integral to our understanding in this section.

In Chapter Twelve, we saw some of the other positive effects that cannabinoids can have for cancer patients, such as inducing appetite, helping to keep body weight steady, preventing nausea and vomiting, and alleviating pain while avoiding opioids, which are highly addictive. We saw that the two pharmaceuticals that are approved by the DEA and FDA for use by the public that also contain cannabinoids simply treat these symptoms though, and we felt it a shame that the anti-cancer qualities of these substances is being so ignored. We saw too that CBD and THC can enhance the effects of traditional cancer treatment methods so that a toxic dose of radiation or chemotherapy is no longer necessary.

In Chapter Thirteen, we looked in depth into various types of cancers and what studies have to say about the interaction of cannabinoids on these different varieties of the malignant cancerous cells. We looked at bladder cancer, brain cancer, and breast cancer in particular.

In Chapter Fourteen, we discussed the various means by which a person might ingest or inhale medical marijuana and cannabinoids. We evaluated and discovered that the extremely concentrated CBD and cannabinoid oils are likely going to be the most effective because of their efficiency in conveying bioavailability of the substance they transport into the body.

Finally, in Chapter Fifteen, we discussed various strains of Cannabis and how they might affect different cancer symptoms and cancer treatment side effects. We looked at overall symptoms, pain, nausea, appetite loss, depression, and fatigue to give you some ideas of how different strains interact with the human body and how they can treat your symptoms.

To conclude, studies indicate that cannabinoids could both enhance the effectiveness of current cancer treatment methods as well as might be a possible treatment option on their own. Each of delta-9 tetrahydrocannabinol (THC) and cannabidiol (CBD), as well as cannabidiolic acid (CBDA), cannabichromene (CBC), and cannabigerol (CBG), has its own method of treating cancer, sometimes more than one. Indeed, in various quantities and mixtures, these substances have been known to be anti-proliferative, anti-metastatic, preventative of angio-genesis, and promoting of apoptotic cell death. Thereby they interrupt the processes of cancer and could help to restrain and manage, if not cure, cancerous tumors and diseases.

CLICK HERE for your FREE BONUS MATERIAL!